ANCIENT ASTRONAUTS

BY SUE HAMILTON

VISIT US AT
WWW.ABDOPUBLISHING.COM

Published by ABDO Publishing Company, 8000 West 78th Street, Suite 310, Edina, Minnesota 55439.
Copyright ©2008 by Abdo Consulting Group, Inc. International copyrights reserved in all countries.
No part of this book may be reproduced in any form without written permission from the publisher.
Abdo & Daughters™ is a trademark and logo of ABDO Publishing Company.

Printed in the United States.

Editor: John Hamilton
Graphic Design: Sue Hamilton
Cover Design: Neil Klinepier
Cover Illustration: Alien, Corbis
Interior Photos and Illustrations: p 4 *Chariots of the Gods?*, courtesy Bantam Books; p 5 Alien, Corbis;
pp 6-7 Sun over the Great Pyramid, Getty; p 9 Woman on Pyramid, Corbis; p 11 Dr. Zahi Hawass,
Getty; p 12 Teotihuacán, © Philip Baird; p 13 Warrior skeletons, Corbis; p 14 (top) UFO, iStockphoto;
(bottom) People at Teotihuacán during spring equinox, AP Images; p 15 People climbing the Pyramid
of the Sun, AP Images; p 16 Starchild skull, courtesy Lloyd Pye; p 17 Starchild illustration, ©2007
RobRoy Menzies; p 18 Scientist conducting radiocarbon dating, AP Images; p 19 Gray alien, iStockphoto;
p 20 Native Americans with flattened heads, North Wind Picture Archives; p 21 Universe man,
Corbis; pp 22-23 Nazca spider, Getty; p 24 Maria Reiche, AP Images; pp 24-25 Nazca hummingbird,
© Philip Baird; p 26 Nazca monkey, AP Images; p 27 Nazca astronaut, Corbis; p 28 (top) Val Camonica
petroglyph, courtesy UNESCO; (middle) Superstition Mountains petroglyph, iStockphoto; (bottom)
Moab man petroglyph, iStockphoto; p 29 Australian cave paintings, Corbis.

Library of Congress Cataloging-in-Publication Data

Hamilton, Sue L., 1959-
   Ancient astronauts / Sue Hamilton.
      p. cm. -- (Unsolved mysteries)
   Includes index.
   ISBN 978-1-59928-833-8
   1. Archaeoastronomy. 2. Extraterrestrial beings. 3. Extraterrestrial anthropology. I. Title.

GN799.A8H36 2007
001.942--dc22
                        2007014552

CONTENTS

ANCIENT ASTRONAUTS: FACT OR FICTION?

Around the world, from Egypt's Great Pyramid to the mysterious, giant-sized drawings of Nazca, Peru, brilliant minds from long ago created incredible structures, sculptures, and artwork. As archeologists shed new light on these artifacts, people wonder how ancient civilizations could possibly have constructed such astonishing works. Some are convinced humankind must have had help from the minds and machines of aliens.

In 1969, Swiss author Erich von Däniken published *Chariots of the Gods?* The author claimed to have "…scientific evidence from five continents that proves that Earth has been visited repeatedly by advanced aliens from other worlds. They left their mark in ancient ruins, lost cities and spaceports…" The author also believed that humankind's ancestors had alien relatives, ancient astronauts from another world. After reading the best-selling book, millions of people around the world began to wonder.

However, most scientists doubt von Däniken's theories. Isaac Asimov, author, historian, and professor of biochemistry, said in his book *Extraterrestrial Civilizations*, "It is a mistake to believe that the ancients were not every bit as intelligent as we. Their technology was more primitive, but their brains were not."

Still, many of these massive monuments would be difficult to reproduce even with today's advanced engineering skills and high-tech equipment. How did the ancients do it? Is it possible they had extraterrestrial help? Some people firmly believe so.

The study of ancient astronauts is called paleo-SETI. Paleo means "ancient times." SETI means "Search for Extraterrestrial Intelligence." One group of paleo-SETI researchers call themselves AAS RA (Archeology, Astronautics and SETI Research Association). Members of this group believe that aliens came to Earth long ago and helped humans learn and develop.

Below: Chariots of the Gods? was published in 1969 by Erich von Däniken. He presented proof that advanced aliens had been on Earth. Many scientists disagreed with his findings.

ERICH VON DÄNIKEN
CHARIOTS OF THE GODS?

MEMORIES OF THE FUTURE—
UNSOLVED MYSTERIES OF THE PAST

Giorgio Tsoukalos, chairman and director of research for the organization, said, "All we do is present evidence, and then we let everybody else make up their own mind."

As new evidence comes to light, scientists and historians continue exploring the most ancient sites around the world, looking deep into the unexplained mystery of ancient astronauts.

Above: Did ancient astronauts help humans build the great monuments, buildings, and art that still stand today?

WHO BUILT EGYPT'S PYRAMIDS?

Egypt's pyramids are world famous. Millions of people flock to see the three most famous pyramids at Giza, near Cairo, Egypt. Of most interest is the Great Pyramid. Egyptologists think this massive structure was created as a tomb for the Egyptian Pharaoh Khufu. If Khufu wanted a tomb that would stand the test of time, he got his wish. Completed around 2500 B.C., the Great Pyramid is one of the Seven Wonders of the Ancient World, and the only structure on the list still standing.

The base of the Great Pyramid covers 13 acres (5.3 hectares) of land. There are 203 layers of granite and limestone blocks stacked in perfect alignment. At 483 feet (147 m), it is the same height as a 40-story building. After it was built, the Great Pyramid remained the world's tallest structure for nearly 4,000 years.

Construction is believed to have taken between 20 and 100 years. Thousands of workers toiled to place approximately 2.3 million stones. Each stone weighed from 1.5 tons (1.3 metric tons) to 70 tons (63.5 metric tons). Simply moving each monstrous block was an engineering miracle. What is even more amazing is that each stone was cut in exacting detail to form a perfect fit.

Facing Page: The Great Pyramid at Giza, near Cairo, Egypt. Some believe that ancient Egyptians had help constructing the massive structure. Egyptologists say that this idea takes credit away from the Egyptians.

Casing stones, shining marble-like white limestone that once covered the Great Pyramid, were also precisely cut. Even today, using laser technology, it would be very difficult to create this precision. Some say the work was done with bronze saws set with diamond cutting points. Others claim the Egyptians needed advanced technology to cut the stones.

Everyone agrees the pyramids are a work of genius. British Egyptologist Sir William Flinders Petrie said the precision was "the finest opticians' work on a scale of acres." Others think the work is too good—too advanced.

Christopher Dunn, an engineer and machine tool manufacturer, examined the tools and some of the artifacts created by the ancient Egyptians. Dunn stated, "They are simply physically incapable of reproducing those artifacts today. So why should I believe they could do so thousands of years ago?" Dunn believes that the Egyptians must have used some type of ultrasonic drill to cut through the solid granite stones. "Today's ultrasonic drills use very high frequency vibration, sound, traveling through a medium."

Could the ancient Egyptians have had such advanced machines back then? Where did the technology come from, and what happened to it? None of these "space age" tools have ever been found, but as Dunn has stated, "Ferrous metal tools would have rusted away by now."

Some researchers wonder if Egyptian engineers had help, maybe even from ancient astronauts. Was it extraterrestrial technology that was used to cut the rock? Did alien visitors use levitation devices to move the giant stones to the Great Pyramid?

Below: Hieroglyphs that resemble modern helicopters, submarines, and gliders from the New Kingdom Temple at Abydos.

Some believe ancient astronauts did assist the Egyptians. Stated Giorgio Tsoukalos of AAS RA, "Technologically speaking we were still primitive and they just gave us a gentle push in the right direction. …they taught us mathematics, astronomy, agriculture and so on and so forth." Tsoukalos believes ancient visitors provided the Egyptians with the knowledge and understanding needed to create their amazing pyramids. As proof, some people look to the hieroglyphs found in the nearby New Kingdom Temple at Abydos. Carvings that appear to be modern helicopters, submarines, and gliders are on a ceiling beam of the 3,000-year-old temple. Could these images prove that ancient astronauts visited Earth?

Above: A woman stands near one of the blocks of an Egyptian pyramid. The Great Pyramid is made up of approximately 2.3 million of these giant stones, each weighing from 1.5 tons (1.3 metric tons) to 70 tons (63.5 metric tons). Some people wonder how the Egyptians could move such great blocks. Did they have help from ancient astronauts? Most scientists say the builders used their own knowledge and muscles, not high-tech equipment.

Most Egyptologists strongly disagree that these carvings are anything but hieroglyphs. They guess that a recarving took place. Hieroglyphs were carved into the temple's sandstone wall. After some time, for whatever the reason, new hieroglyphs were carved, but the old hieroglyphs were only partially erased from the sandstone wall. Some of the old characters remained behind the new characters. As the recarved hieroglyphs aged, the old and new characters blurred into one another. People viewing the images today think the pictures look like modern modes of transportation. However, any resemblance to today's machines is simply a coincidence.

Dr. Zahi Hawass, Egypt's secretary general of the Supreme Council of Antiquities, uses the term "pyramidiots" in referring to people who believe the ancient Egyptians had extraterrestrial help to build the pyramids. Hawass and many other scientists and historians feel this idea takes credit away from the ancient Egyptians. In a 1997 interview, Hawass stated, "Not a single piece of material culture–not a single object–has been found at Giza that can be interpreted to come from a lost civilization. Instead we find an abundance of tombs, bodies, ancient boats, hieroglyphic inscriptions, pottery, bakeries and so on, from the Egyptian culture of the 4th Dynasty, about 2,500 B.C. Theories and speculations about a lost civilization seem to excite people more than the discovery of a culture that we actually find at Giza and elsewhere in Egypt, the culture of the Egyptians of whose existence we are certain. It was a great culture. Why do people need to look for another?"

Most archeologists and Egyptologists agree with Dr. Hawass. Kenneth Feder, anthropologist and author of *Frauds, Myths, and Mysteries: Science and Pseudoscience in Archaeology*, says, "Anyone who is interested can visit Egypt and see paintings on the walls of ancient buildings and tombs that depict exactly how large construction projects were carried out. To say that the ancient Egyptians could not have built the pyramids would be like saying Americans could not have built the Empire State Building in New York City even though there are photographs of its construction."

Still, the amazing accomplishments of the ancient Egyptians continue to be questioned. Did ancient astronauts help them? The mystery remains.

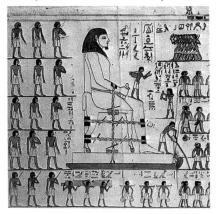

Below: The ancient Egyptians moved heavy objects using a system of ropes and sleds. This illustration shows the statue of Djehutihotep being moved in just such a way.

Above: Dr. Zahi Hawass, Egypt's secretary general of the Supreme Council of Antiquities, uses the term "pyramidiots" to refer to people who believe the ancient Egyptians had extraterrestrial help building the pyramids.

PYRAMIDS OR LANDING PADS?

Above: Some people wonder if Teotihuacán's great pyramid, the Pyramid of the Sun, was once a landing pad for ancient astronauts.

More than 1,800 years ago, the largest city in the Americas was Teotihuacán, located near present-day Mexico City, Mexico. With its soaring pyramids, temples, and sprawling plazas, this bustling metropolis was home to as many as 200,000 people. The city's largest pyramid, the Pyramid of the Sun, is a whopping 705 feet (215 m) long on each side. It stands 212 feet (65 m) tall. Today it is the third-largest pyramid in the world.

Archaeologists don't know exactly who built this amazing city. No written documents exist to explain how it came to be or why it was eventually abandoned. However, many people guess that the Pyramid of the Sun and the other structures in the great city were built for religious ceremonies.

The graves of a number of people have been found near the base of the pyramids. The clothing and artifacts found with the skeletons show that these people were not from Teotihuacán. Researchers guess that these unlucky outsiders were probably captured warriors or slaves who were sacrificed in religious ceremonies on the pyramids, then buried nearby.

Some researchers wonder if the city and the pyramids arose for reasons other than religion. One idea is that local people built the enormous structures as landing pads for ancient astronauts. Some suggest that Teotihuacán's name, which means "place of the gods," shows that the city was planned and created by extraterrestrial travelers. Aliens with advanced knowledge and skills would certainly have seemed godlike to the people living there at the time.

Below: A row of nine skeletons of sacrificed warriors excavated from the "Street of the Dead" in Teotihuacán.

Above: A UFO lands on a hilltop. *Below:* People welcome the spring equinox by raising their hands at sunrise from the top of the Pyramid of the Sun.

Additionally, some people believe that Teotihuacán was built for the mining and collection of gold for ancient astronauts. The theory is that the extraterrestrials' planet had experienced some type of global warming. They needed gold to help save their home world. They discovered that Earth contained a lot of gold. With so many people living in Teotihuacán, there were many humans available to mine the gold. The aliens used their time here to educate humans, teaching and assisting the people to build the pyramids at "the place of the gods."

The only problem with this theory is that there is no gold in or near Teotihuacán. Kim Goldsmith, a long-time archeologist of the area, stated in a National Geographic documentary, "There was no gold here. There was no gold anywhere in this area to start off with. There were no gold mines and they were not bringing gold from anywhere during Teotihuacán times." Still, the theories continue to be researched and debated.

Below: Hundreds of people climb the Pyramid of the Sun, the third biggest pyramid on Earth.

THE STARCHILD SKULL

Some researchers believe that ancient astronauts not only taught humans, but also stayed on Earth, married, and had children. A human-alien hybrid race may have been created, which explains how people could leap from simple caveman-like beings to master architects and engineers in a period of a few thousand years. Some even say proof may be available from a mysterious skull found in Mexico.

In 1930, a teenage girl found an old mine tunnel in northwest Mexico. Her curiosity caused her to wander inside, where she soon made a grisly discovery: bones. Sweeping away some of the surrounding dirt, the girl realized she had uncovered an ancient grave. In front of her lay two skeletons. One was an adult. The other set of bones appeared to be those of a child. But there was something very odd about the smaller skull.

Many years passed, and the mysterious skull changed owners several times. Eventually it came into the possession of Melanie Young, a nurse from Texas who worked with newborn children. As a medical professional, Young was fascinated with the unusual shape of the skull, with its larger-than-normal cranial area and very shallow eye sockets. Aside from its odd shape, the skull was also

Below: The Starchild skull.

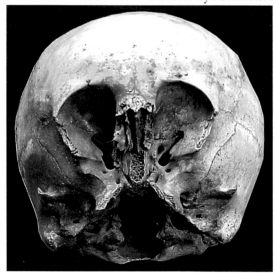

very light. Young estimated that it weighed about half as much as a normal child's skull. She began to wonder if the skull was even human.

In February 1999, Young turned the skull over to Lloyd Pye, an author and researcher in what he calls "alternative knowledge." Pye has studied the theory that ancient astronauts mated with early humans to create an alien/human hybrid, whose ancestors walk among us even today. Pye says he researches the alien/ human hybrid theory using science, as well as human history and evolution.

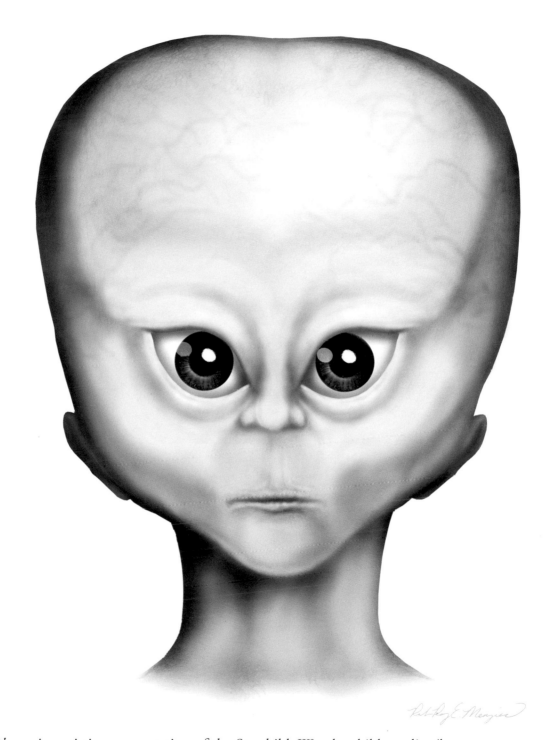

Above: An artistic representation of the Starchild. Was the child an alien/human hybrid or simply a child suffering from some type of physical deformity or serious medical condition?

Pye decided to bring the skull to a large UFO conference in Nevada. He showed it to several people. Many wondered if they were looking at an actual skull of a "gray," one of the most commonly reported human-like aliens. This extraterrestrial is described as being short and thin with grayish skin, a large head, teardrop-shaped eyes, and a small nose, mouth, and ears. Did Pye obtain actual proof of an alien/human life form? The skull was nicknamed the "Starchild."

Pye decided to have the Starchild scientifically tested. Through radiocarbon dating, the skull was estimated to be about 900 years

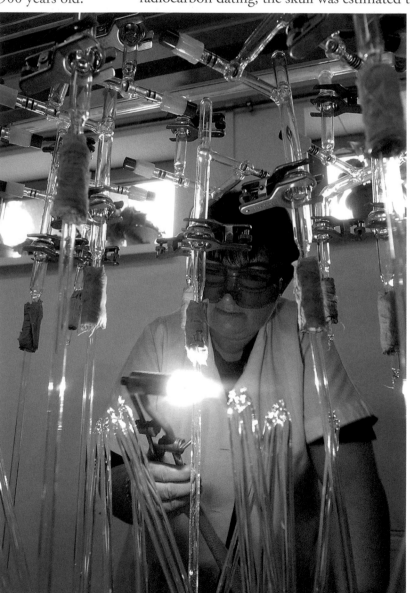

Below: A lab technician conducts radiocarbon dating. Through this process, the Starchild skull was found to be about 900 years old.

old. Later, in separate tests, the bone material was found to contain both an odd red substance and some unknown fibers.

Pye's next step was to test the skull's DNA, the genetic instructions of all living things. DNA is like a blueprint, or an instruction manual, that is in every cell of an organism's body. Unique traits—hair color, skin color, sex, height—are all contained in the DNA instructions. Unfortunately, it is very difficult to extract DNA from something old. Over time, the cells break down, which makes it hard to find intact DNA molecules.

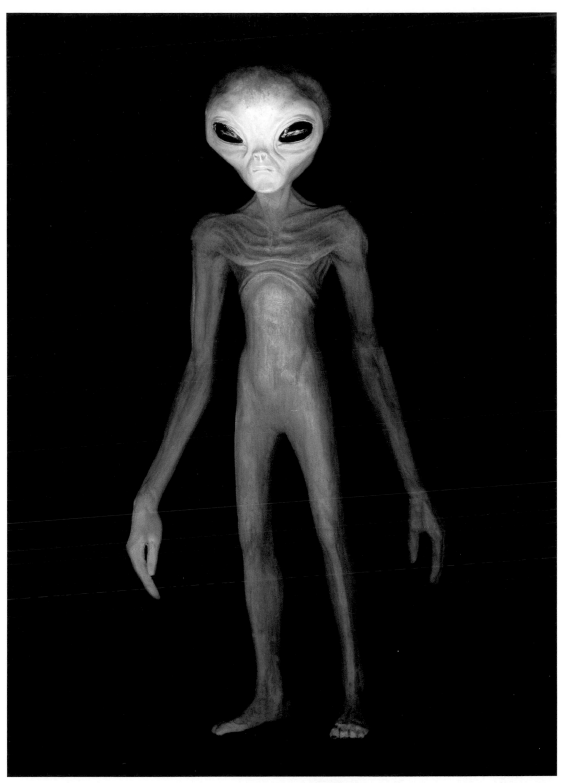

Above: A "gray" is one of the most commonly reported aliens.

There was another problem that Pye worried about: Do aliens even have DNA? Since the laboratory scientists were looking for human DNA in the Starchild skull, would they miss possible evidence of alien genetic material?

When the results came back, the lab confirmed that the skull belonged to a child, about five years old, whose mother was human. The first test indicated that the child was a young boy, but that finding was later questioned. To date, scientists are uncertain if the Starchild was a boy or a girl. Also, the lab tests were unable to determine anything about the Starchild's father. Is it possible the missing father was an alien?

Dr. William Rodriguez, a forensic anthropologist, was called in to give his opinion of the Starchild skull. A forensic anthropologist is someone who uses scientific methods to identify skeletal remains. He thought that the skull showed a simple human deformity.

Dr. Steven Novella, a medical doctor, provided another opinion. He believed the Starchild skull's appearance was caused by a medical condition known as hydrocephalus, sometimes called "water on the brain." In this condition, a blockage prevents cerebrospinal fluid from flowing out of the brain. The fluid builds up inside the skull. Dr. Novella said, "Because in young children the bones of the skull have not yet fused together, the skull is free to enlarge to accommodate this buildup of fluid." This condition would cause an enlarged skull. The child's large head would look like a gray alien's head. Lloyd Pye and his supporters disagree with these findings. They cite the mysterious red substance and strange fibers found inside the bone tissue, as well as the skull's overall abnormal shape.

Pye said, "The answer seems to be that its genes told it to grow this way and every compensation that needed to be made was made to accommodate all of these very significant changes that are simply not seen in normal human beings."

Below: Some Native American tribes used cradleboards to flatten their children's heads. It's possible that the Starchild skull looks the way it does because of this process.

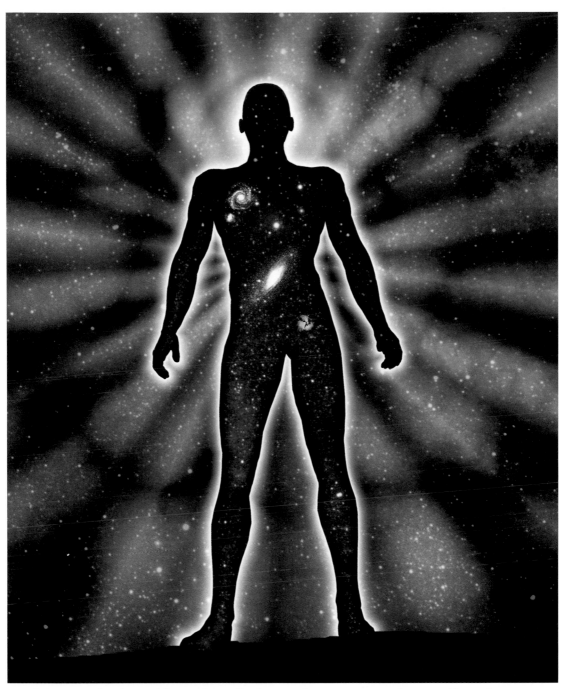

Science today is unable to give a final, complete answer to the history of the Starchild. Was it a human child suffering from an unusual medical condition, or was it the product of alien/human mating? As scientific processes are perfected, perhaps in the future we'll discover the answer. For today, the Starchild remains a mystery.

Above: Are some people really alien/human hybrids?

THE NAZCA LINES OF PERU

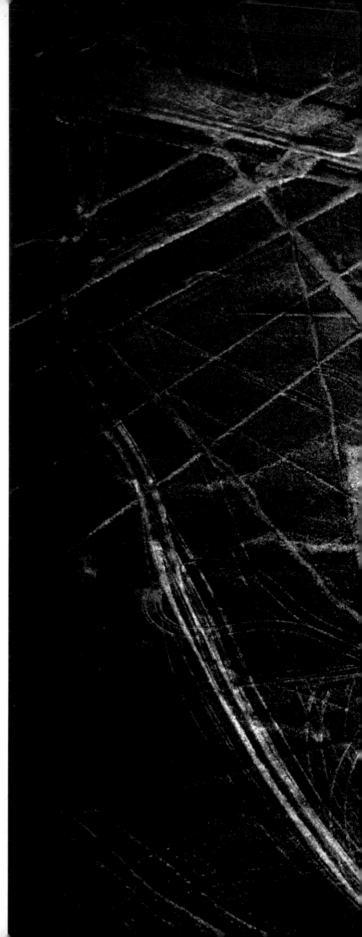

In 1926, archaeologists Toribio Mejia and Alfred Kroeber discovered what they thought were shallow irrigation ditches on a high, arid plateau of the Nazca Desert, in Peru, South America. The scientists believed that the ancient people of Nazca built the irrigation ditches to distribute water across the parched land. Nazca is a flat, dry, windless place that receives less than one inch (20 mm) of rain in an entire year, perfect conditions for preserving archaeological ruins.

When airplanes began flying over the area several years later, however, pilots and passengers got a much different view of the scene. From the air, the simple ruts in the ground transformed into easily recognizable figures. Visible only to airborne observers are more than 300 giant-sized designs spread over the desert, including a monkey, a spider, a hummingbird, and even what appears to be a 100-foot (30-m) –tall astronaut. The Nazca lines also include abstract figures, both straight and patterned.

Right: Nazca lines and the spider image as seen from the air.

Called geoglyphs, or drawings on the ground, the Nazca lines were created by native people more than 2,000 years ago. They formed the lines by brushing aside the dark pebbles on the desert floor to reach the light-colored sand below. The patterns cover an area of more than 193 square miles (500 sq km). How did the Nazca people accomplish this feat, and why? Did they have a way to fly into the air? Or were alien ships hovering above, instructing the Nazca people how to create the geoglyphs?

Anthropologist Kenneth Feder believes the Nazca created these geoglyphs without any help from ancient astronauts. In *Frauds, Myths, and Mysteries: Science and Pseudoscience in Archaeology*, he states, "They [the figures and lines] are remarkable achievements because of their great size, but certainly not beyond the capabilities of prehistoric people. Remember, these drawings were not carved into solid rock with extraterrestrial lasers; they were not paved over with some mysterious substance from another world. They were, in essence, "swept" into existence."

Some people believe the long, straight lines were landing strips for ancient astronauts and their spacecrafts. Viewed from above, the lines certainly look like they belong on a modern airport. In the 1968 book *Chariots of the Gods?*, author Erich von Däniken stated, "Seen from the air, the clear-cut impression that the 37-mile-long plain of Nazca made on *me* was that of an airfield… What is wrong with the idea that the lines were laid out to say to the 'gods': 'Land here! Everything has been prepared as you ordered.'" However, most scientists strongly disagree with von Däniken's theory of ancient astronauts visiting Nazca.

Below: Maria Reiche, a German mathematician and astronomer, spent more than 50 years studying the Nazca lines. She became known as the protector of the ancient geoglyphs until her death at the age of 95 in June 1998.

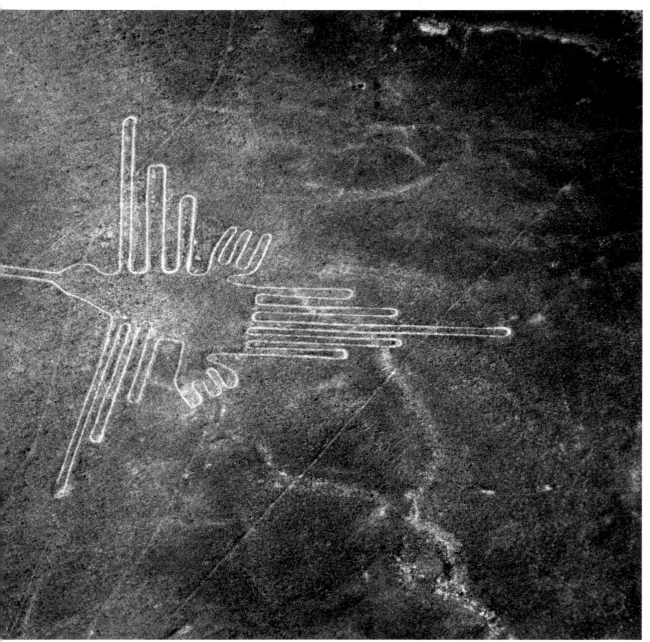

In 1941, Paul Kosok, an American historian, and Maria Reiche, a German mathematician and astronomer, began studying the Nazca lines. They discovered that some of the lines marked the changing seasons, such as the summer and winter solstice, the longest and shortest days of the year. They began to believe the lines might be some type of giant calendar.

Above: In this hummingbird image a long path leads into and out of the design. Is this a ceremonial walking path?

Above: This giant curly-tailed monkey is big enough to fill a football stadium. The monkey has four fingers on one hand and three on the other. Is this an important clue or just an accidental mistake by the Nazca creators?

Twenty-seven years later, in 1968, the same year that von Däniken's ancient astronauts theory was published, astronomer Gerald Hawkins arrived in Nazca. He discovered that while some of the lines did mark specific sun, moon, and star positions, many of the other lines were haphazardly placed. It seemed that researchers were right back where they started, questioning why the lines were ever created.

In the 1980s, anthropologists Anthony Aveni and Helaine Silverman began working with Tom Zuidema, an expert on Inca culture. Knowing that the Incas had ruled much of Peru before the Spanish arrived in the 16th century, the three researchers wondered if some of the Incan beliefs were shared by the Nazca civilization. Zuidema knew that the Incas would often use straight lines flowing out from a center point in their religious ceremonies. Perhaps the Nazca believed similarly? This also reinforced the idea that religion might have played a part in the huge artistic drawings.

The giant images, lines, spirals, and zigzags seemed designed for religious ceremonies. Stated Anthony Aveni, "I think the lines were made to walk upon; they were pathways." Aveni and other anthropologists believe that the patterns may have been of ancient animal gods, and that the Nazca built a path up to the design. Like a starting point on an Etch-a-Sketch drawing toy, a long path connects the starting point with the actual design. A close look at the hummingbird shows how the path starts off at one point, merges into the design, and then leads back off. Was it a ceremonial walking path?

Some people disagree. They point out that one of the images looks just like an alien astronaut. Perhaps aliens visited Nazca and then returned to their home planet in the far-away reaches of space. One idea is that the giant runways and images were an invitation by the Nazca people for the extraterrestrials to return.

Studies of these fascinating lines and images continue to yield guesses and ideas, but the final answer to the mystery is yet undiscovered.

Above: Could this 100-foot (30-m) -tall hillside Nazca drawing be an ancient astronaut?

CAVE AND ROCK PAINTINGS

Around the world ancient cave and rock paintings have been discovered that appear to be pictures of aliens. Could these be images of ancient astronauts, recorded thousands of years earlier by our ancestors? Or are they simply early drawings, perhaps of people or gods? Have we just imagined them to be something more than what they really are?

In Australia, aboriginal art depicts some very interesting characters. The Wandjina, or spirit guardians, were created around 3000 B.C. in western Australia. Reportedly, the guardians watched over the land. The artistic look of some of the spirit guardians is that of an alien with a large head and big eyes. Several different images have been found near Victoria, Australia. These include the Lightning Brothers Tjabuinji and Jagtjadbulla, as well as antenna figures and a fascinating helmeted figure, which appears to be wearing a spacesuit and gloves.

Petroglyphs found in the United States show some unique creations as well. In California, Arizona, Wyoming, New Mexico, Utah, and Arkansas are rock and cave drawings that don't seem to quite fit with the native culture who produced the art. Did early Native Americans see extraterrestrials? Or are these drawings just images of gods or spirits, or perhaps religious leaders wearing special headdresses?

Rock drawings in Val Camonica, a valley in the Lombardy region near Milan, Italy, show interesting scenes. Some say the petroglyphs drawn by the Camunni tribe some 10,000 years ago show ancient astronauts. Others say the drawings are simply pictures of hunters and tribe ceremonies.

Left top: A rock drawing from Val Camonica, Italy. Some believe the petroglyph shows ancient astronauts wearing helmets.
Left middle: A petroglyph from the Superstition Mountains in Arizona.
Left bottom: The Moab Man, from Moab, Utah. Is he wearing a headdress, or does he have antenna coming from his head?

Did ancient astronauts really visit our planet? Wernher von Braun, a rocket scientist who helped create the Saturn V rocket that brought the Apollo 11 crew to land on the moon in 1969, once said: "Our sun is one of 100 billion stars in our galaxy. Our galaxy is one of billions of galaxies populating the universe. It would be the height of presumption to think that we are the only living things in that enormous immensity."

Perhaps ancient humans left hints to us about the extraterrestrials that visited our planet long ago. Perhaps these ideas are all imaginative theories. Without scientific proof, the idea of ancient astronauts remains an unsolved mystery.

Above: Aboriginal cave paintings near the Sale River in Kimberley, Australia.

GLOSSARY

ANTHROPOLOGIST

A person who studies the origin and history of people in a specific area and time. Anthropologists try to learn about a people's behavior, life, speech, homes, and overall culture.

ARCHAEOLOGIST

A person who searches for, uncovers, and studies artifacts from the past in order to learn how people of ancient societies once lived. Archaeologists try to discover information about an area's people, such as the foods they ate, the kind of homes they lived in, who was in charge, and what was important to them.

ARTIFACTS

Objects made or owned by a person or group of people. Often of historical interest, artifacts can be something worn, like clothing or jewelry, or something used, such as dishes or tools, or something created by a culture, such as art.

DNA

DNA is short for the scientific term Deoxyribonucleic Acid. In living things, DNA is the material inside the center of every cell that forms genes. This material is inherited from a person's or other living thing's parents. Except for identical twins, each person's DNA is unique to that person. Identical twins have identical DNA.

EGYPTOLOGIST

A person who studies ancient Egypt, its people, artifacts, and culture. A highly developed ancient Egyptian society existed for thousands of years, providing much to be discovered and explored.

EXTRATERRESTRIAL

Something that comes from outside the earth or its atmosphere.

GALAXY

A system of millions, or even hundreds of billions, of stars and planets clustered together in a distinct shape, like a spiral or ellipse. Our Earth is located within the Milky Way Galaxy.

GENES

Genetic information that is inherited from a parent to their offspring. Many characteristics of a person's body are determined by what that person inherits from their father and mother. For example, tall parents are likely to pass this gene on, so their children will probably also be tall.

HIEROGLYPHS

Characters in a writing system, especially those used by ancient Egyptians. Hieroglyphs are mostly pictorial, not alphabetical, which makes them difficult for outsiders to understand. The study of ancient Egypt was greatly advanced when hieroglyphs were deciphered, or translated.

LEVITATION

To rise and hover in the air as though weightless, especially through the use of supernatural or magical powers. On Earth, where gravity holds all things down, objects can only appear to levitate. An outside force, such as blasts of air, are needed to cause an object to float up against gravity.

MONUMENT

A statue, building, or other structure created to remember a great person, group of people, or event.

PETROGLYPH

A carving or drawing on rocks, usually by prehistoric people, for artistic or religious purposes.

SACRIFICE

The act of killing an animal or person as an offering to a god or gods, or other important figure.

SEVEN WONDERS OF THE ANCIENT WORLD

The seven most amazing man-made creations produced in ancient times. The Seven Wonders of the Ancient World include the Great Pyramid of Giza, the Hanging Gardens of Babylon, the Statue of Zeus at Olympia, the Temple of Artemis, the Mausoleum of Maussollos, the Colossus of Rhodes, and the Lighthouse of Alexandria.

THEORY

An idea or group of ideas that explains something. A theory is usually based on taking what is known about something—the facts—and combining them into a logical idea.

UFO

An Unidentified Flying Object. A UFO is typically some kind of alien craft, such as a flying saucer.

INDEX